The Science Experiment

Charles Harriman

Charles Harriman Publishing

The Science Experiment

Published in the year 2015 by
Charles Harriman Publishing
Ferndale, Washington

ISBN-13: 978-0-9882408-2-7

Astronomy Photograph by ***NASA*** with image
from the *Hubble News Release Image Archive.*

Contents

Intro

The environment between two spinning objects may have dynamic interaction.

Two nested spinning objects may attract, repel, or remain depending upon the environmental interaction that occurs between the two nested spinning objects.

If destructive interaction occurs between two nested spinning objects, then they may remain nested with **dynamic interactive stability!**

What happens if constructive environmental interaction occurs between two nested spinning objects?

What happens if an overwhelming amount of destructive environmental interaction occurs between two nested spinning objects?

Set-up

The Science Experiment
Cake Pan Spin & Object Spin Experiment

Materials:

• A round cake pan (9 to 24 inches in diameter)
• A top
• A turntable and bowl may be used to spin the cake pan (secure the bowl to the disc of the turntable, and then set the cake pan onto the bowl).

Notes:

* All experiments in this book are to be done carefully and with adult supervision. The author of this book is not liable for any injuries or damages from doing these experiments.
* This guide includes four other *Flow Theory* experiments that support the ***Cake Pan Spin & Object Spin Experiment***.
* A procedure, diagram, observation, and *Flow Theory* interpretation are given for each experiment. The book *Flow Theory* by Charles Harriman is a scientific geometry and a new Theory Of Everything.
* *Flow Theory* assumes that an environment of infinitely smaller matter exists at all scales of the universe.
* *Flow Theory* assumes that low pressure environments exist beyond the atmosphere of the Earth and also between chemical molecules.
* View a video slide show of the ***Cake Pan Spin & Object Spin Experiment*** online. Search: *Flow Theory Experiments.*

The Science Experiment

Cake Pan Spin & Object Spin Experiment

Procedure:

1) Spin a round cake pan, and then spin a top with the same spin direction as the cake pan at the center of the spinning cake pan.
(A turntable and bowl may be used to spin the cake pan.)
2) Spin a round cake pan, and then spin a top with a ***counter*** spin direction to the cake pan at the center of the spinning cake pan.

Diagram:

1)

2)

Observations:

• A spinning object, at the center of a spinning cake pan, that is spinning in the same circular direction as the cake pan may move in an arc toward and following the circumference of the spinning cake pan.
• A spinning object, at the center of a spinning cake pan, that is spinning in a **counter** circular direction to the cake pan may remain spinning at the center of the spinning cake pan.

Interpretation:

According to *Flow Theory*:

In an environment of air with spinning objects and flow systems that are larger than the molecules of air:

• The environment inside and outside of the circumference of a spinning object or circular flow system may flow toward the circumference as the environment follows the inside and outside circular counter flows that are generated by the circumference (see: ***Cake Pan Spin and Objects Experiment***).

• A smaller spinning object that is nested within and is spinning in the same circular direction as a larger spinning object has air flowing toward its circumference that interacts with **constructive interaction** (flows blend) with air flowing toward the circumference of the larger spinning object. The smaller spinning object may then move with the air flowing toward the circumference of the larger spinning object.

• A smaller spinning object that is nested within and is spinning in a **counter** circular direction to a larger spinning object has air flowing toward its circumference that interacts with **destructive interaction** (flows collide) with air flowing toward the circumference of the larger spinning object. The smaller counter spinning object may then remain spinning within the center region of the larger spinning object with **dynamic interactive stability!**

According to *Flow Theory*:

• A smaller spinning object and a larger spinning object in an environment of **low resistance** (such as beyond the atmosphere of the Earth) may both generate outward extending and inward extending circular counter flows, and these circular flows may generate outward extending and inward extending circular counter flows, and so forth.

* *Flow Theory* assumes that an environment of infinitely smaller matter exists at all scales of the universe.

* *Flow Theory* assumes that low pressure environments exist beyond the atmosphere of the Earth and also between chemical molecules.

• Outward extending circular flows generated by the larger spinning object may revolve around the larger spinning object. If the smaller spinning object is nested within the center region of an outward extending circular flow generated by the larger spinning object, then outward extending circular flows generated by the smaller spinning object may interact with **destructive interaction** (flows collide) with inward extending circular flows generated by the circular flow in which the smaller spinning object is nested. The smaller spinning object may then remain spinning within the center region of a circular flow that is generated by and revolves around the larger spinning object, and **dynamic interactive stability** is achieved (without using the concept of gravity) as with a moon orbiting a planet or a planet orbiting a star! 4

Earth & Moon

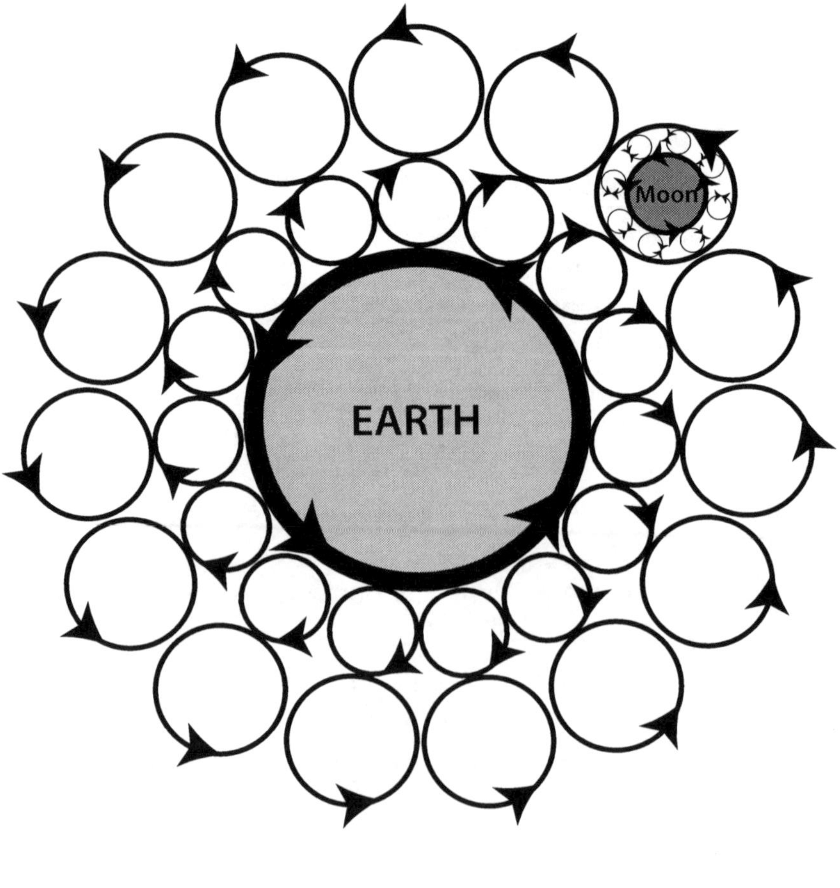

Inner solar system of the Sun

(not to scale)

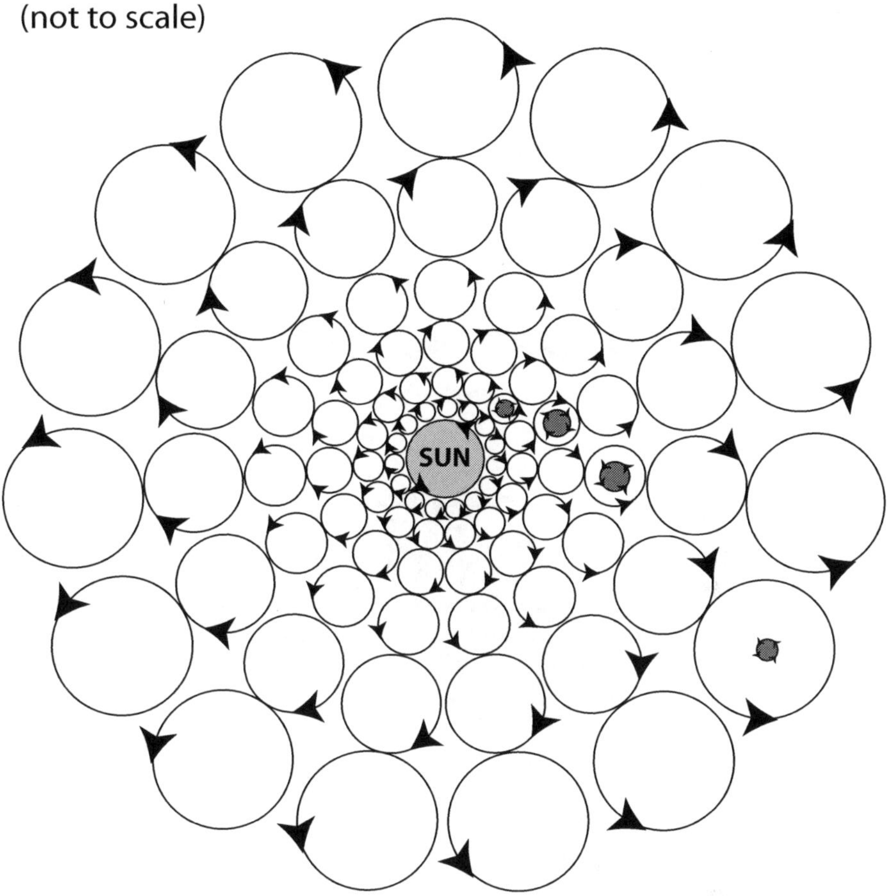

According to *Flow Theory*:

• As the Sun spins in a counterclockwise direction, the Planets of our Solar System spin in the same direction with the circular flow in which they are nested: Mercury (1st planet from the Sun) spins in a counterclockwise direction, Venus (2nd planet) spins in a clockwise direction, Earth (3rd planet) spins in a counterclockwise direction, Mars (4th planet) spins in a counterclockwise direction. Planets farther away from the Sun are in lower pressure environments than planets closer to the Sun, so planets farther away from the Sun may be nested in larger circular flows with faster spin velocities.

• Unperceived flows may circulate between rounds of circular flows (alternating circular directions), and rounds of circular flows may revolve about the Sun at different velocities.

Flow Experiment

Procedure:

Move a flat stick (held vertical) across the surface of water along a straight path with the flat sides of the stick facing the front and back of the stick's direction of motion.

Diagram:

Observations:

- Circular flows of the water occur at each side of a moving stick in the circular direction from the front of the stick around to the sides and back of the stick.
- The environment near the back of the moving stick follows the moving stick.

Interpretation:

According to *Flow Theory*:

• An object in motion or a flow may cause the environment in which it moves to flow in circular patterns in the direction from the front of the object around toward the sides and back of the object (from a higher pressure front to a lower pressure back), and the environment follows the lower pressure back.

• Resistance of an environment affects the formation of circular flows which are generated by a moving object or flow.

• In an environment of greater resistance, a circular flow contracts (as with a gas) and also decreases its circular flow velocity.

• In an environment of lower resistance, a circular flow expands (as with a gas) and also increases its circular flow velocity.

• A circular flow may generate circular counter flows and so forth forming Particle~Waves. Successive circular counter flows form the dual particle and wave nature of light and other Particle~Waves that occur in low resistance environments that exist beyond the atmosphere of the Earth and also between chemical molecules.

Particle~Wave

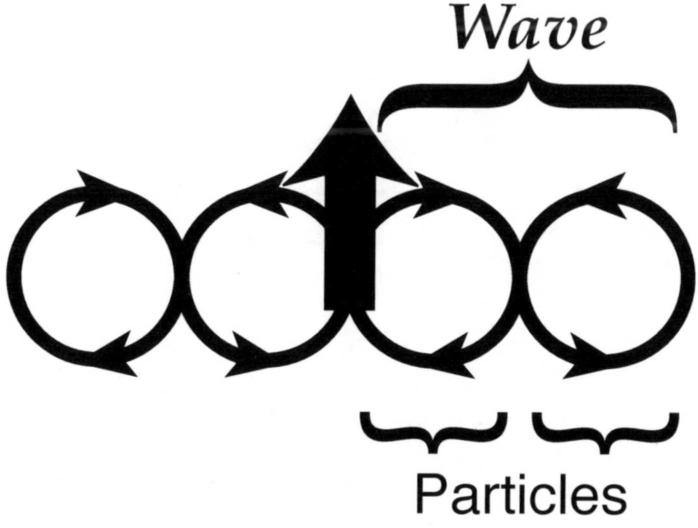

Circular Flow Experiment

Procedure:

Move a flat stick (held vertical) across the surface of water in a **repeating circular pattern** with the flat sides of the stick facing the front and back of the stick's direction of motion.

Diagram:

Observations:

- Circular flows of the water occur at each side of the moving stick in the circular direction from the front of the stick around to the sides and back of the stick.
- These circular flows of the water occur both outside and inside of the stick's repeating circular path.

Interpretation:

According to *Flow Theory*:

• In an environment of **low** resistance, successive circular counter flows may flow across the diameter inside of a circular flow system.

• An **odd** number of diameter circular counter flows may have flows in the **same** circular directions that interact with **constructive** interaction (flows blend) to create a **low** pressure center region as the environment at the center region follows the circular flows that return to the circumference flow.

Low Pressure Circular Flow System

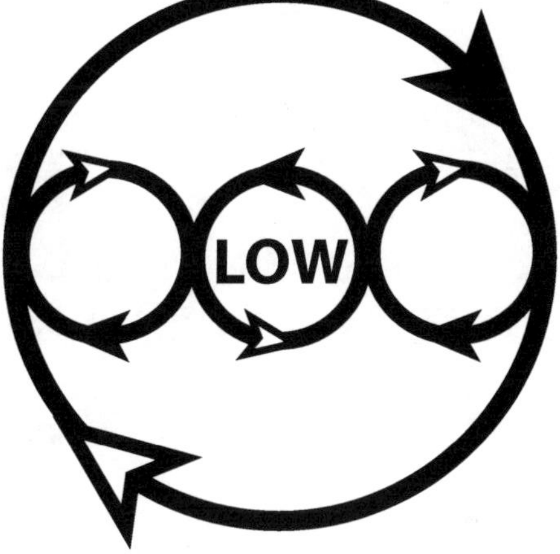

• An **even** number of diameter circular counter flows may have flows in **counter** circular directions that interact with **destructive** interaction to create a **high** pressure center region as the circular counter flows **collide** and the environment at the center region does **not** go to the circumference flow.

High Pressure Circular Flow System

According to *Flow Theory*:

• A Circular Flow System with a **low** pressure center region may form a Top Torus Flow and a Bottom Torus Flow. The environment near the high pressure fronts of the circumference flow flows around above and around below the circumference flow and **inward** toward the low pressure center region (Black Hole) and then to the circumference of the Circular Flow System.

Low Pressure
Circular & Double Torus Flow System

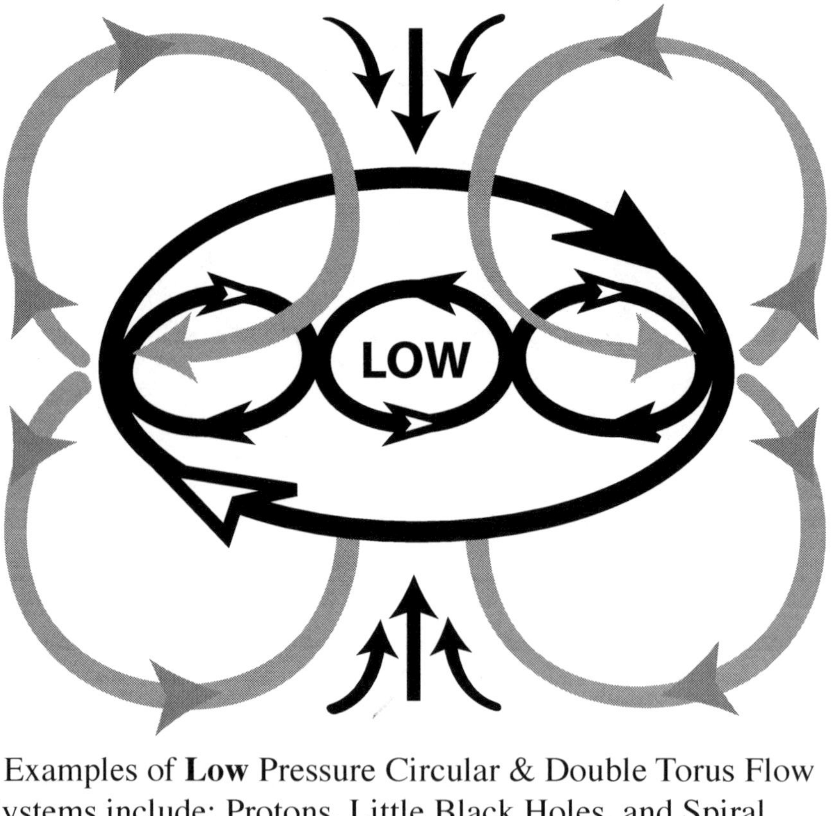

• Examples of **Low** Pressure Circular & Double Torus Flow Systems include: Protons, Little Black Holes, and Spiral Galaxies With Black Holes.

• Torus Flows of Circular & Double Torus Flow Systems may have more curving, twisting, or angular flow patterns than depicted in the graphic.

• Spiral flows of matter (not depicted) may occur around the circumference flow (as with a Spiral Galaxy with a Black Hole) from a lower pressure to a low pressure environment.

According to *Flow Theory*:

• A Circular Flow System with a **high** pressure center region may form a Top Torus Flow and a Bottom Torus Flow. Matter *from* the high pressure center region of the Circular Flow System flows **outward** around above and around below the circumference flow and flows with the environment toward the low pressure sides and backs of the circumference flow of the Circular Flow System.

High Pressure
Circular & Double Torus Flow System

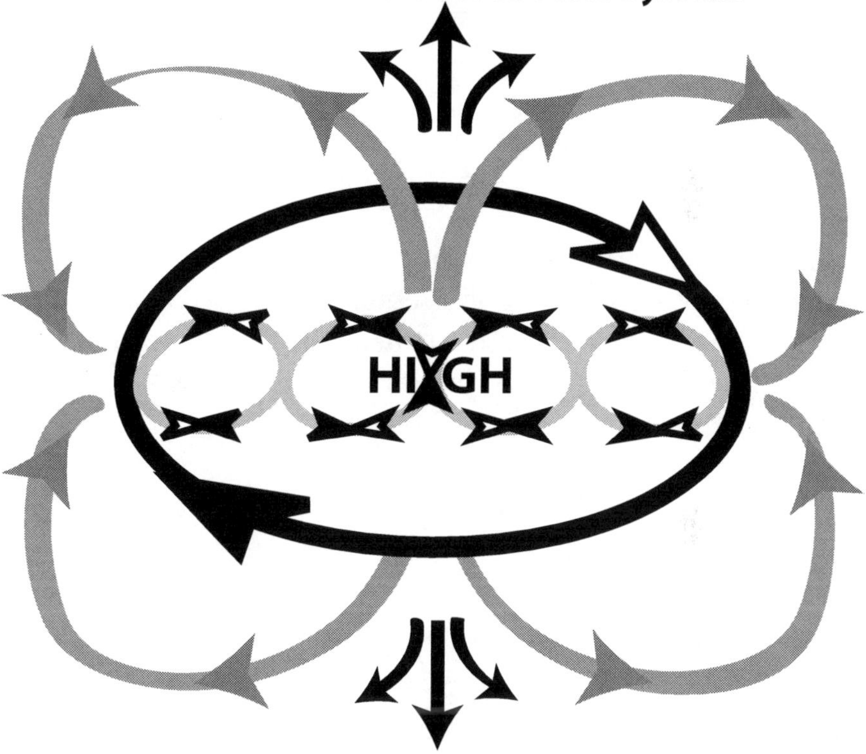

• Examples of **High** Pressure Circular & Double Torus Flow Systems include: Electrons, Stars, and Lenticular Galaxies.

• A disc of revolving circular flows with planets and other matter (not depicted) may occur outside the circumference flow (see: interpretation of Cake Pan Spin & Object Spin Experiment-Inner Solar System of the Sun) from a high pressure to a low pressure environment (with expanding circular counter flows in this case, and with contracting circular counter flows for the lower to low pressure case).

According to *Flow Theory*:

- Protons are **Low** Pressure Circular & Double Torus Flow Systems, and Electrons are **High** Pressure Circular & Double Torus Flow Systems.
- Protons and Electrons are able to expand or contract and increase or decrease their masses according to the amount of flow with their flow systems.

Proton

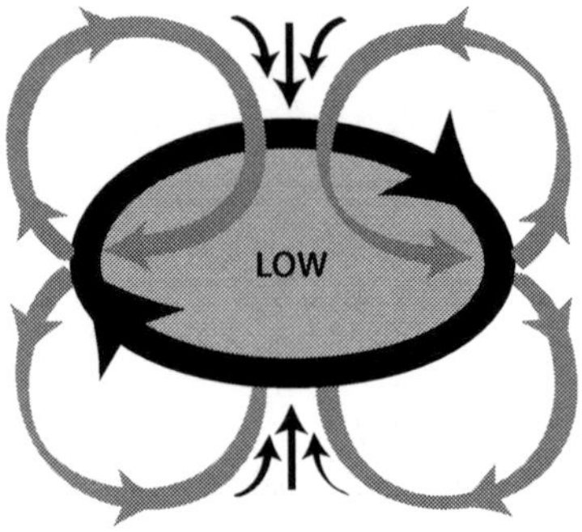

Electron

- Inpouring Low Pressure Circular & Double Torus Flow Systems and outpouring High Pressure Circular & Double Torus Flow Systems may align spins, attract, interact and bond in low resistance environments that exist beyond the atmosphere of the Earth and also between chemical molecules.
- A Low Pressure Proton and a High Pressure Electron may align spins, attract, interact and bond to form the chemical element Hydrogen.
- **Dynamic interactive stability** may be achieved with a Proton and an Electron as their Torus flows interact with destructive interaction.

Hydrogen

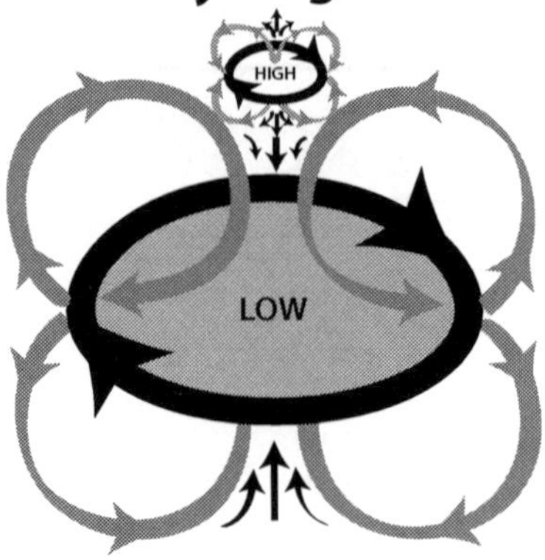

According to *Flow Theory*:

• Two Low Pressure Protons and two High Pressure Electrons may align spins, attract, interact and bend around and bond to form a loop of alternating Protons and Electrons (H-2) with **dipolar** flow.

• **Dynamic interactive stability** is achieved with the loop of Protons and Electrons as their Torus flows interact with destructive interaction.

H-2

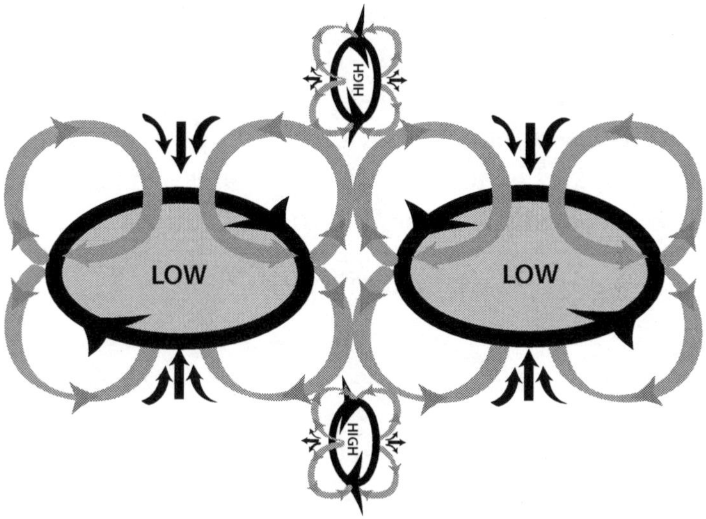

• Protons with two poles of inpouring flow and Electrons with two poles of outpouring flow both have electric qualities, and chemical elements with dipolar flow have magnetic qualities.

H-2 showing Dipolar Flow

According to *Flow Theory*:

• Four Low Pressure Protons and four High Pressure Electrons may align spins, attract, interact and bend around and bond to form a loop of alternating Protons and Electrons with dipolar flow.

• **Dynamic interactive stability** is achieved with the loop of Protons and Electrons as their Torus flows interact with destructive interaction.

• The chemical element Helium (H-4) is a loop of 4 Proton-Electron pairs with dipolar flow (neutrons decay into Protons and Electrons (Malley)).

• The chemical elements are groups of loops of Proton-Electron pairs, and isotopes of the chemical elements are formed by the substitution of H-3 for H-2 and some H-4.

Helium (H-4)

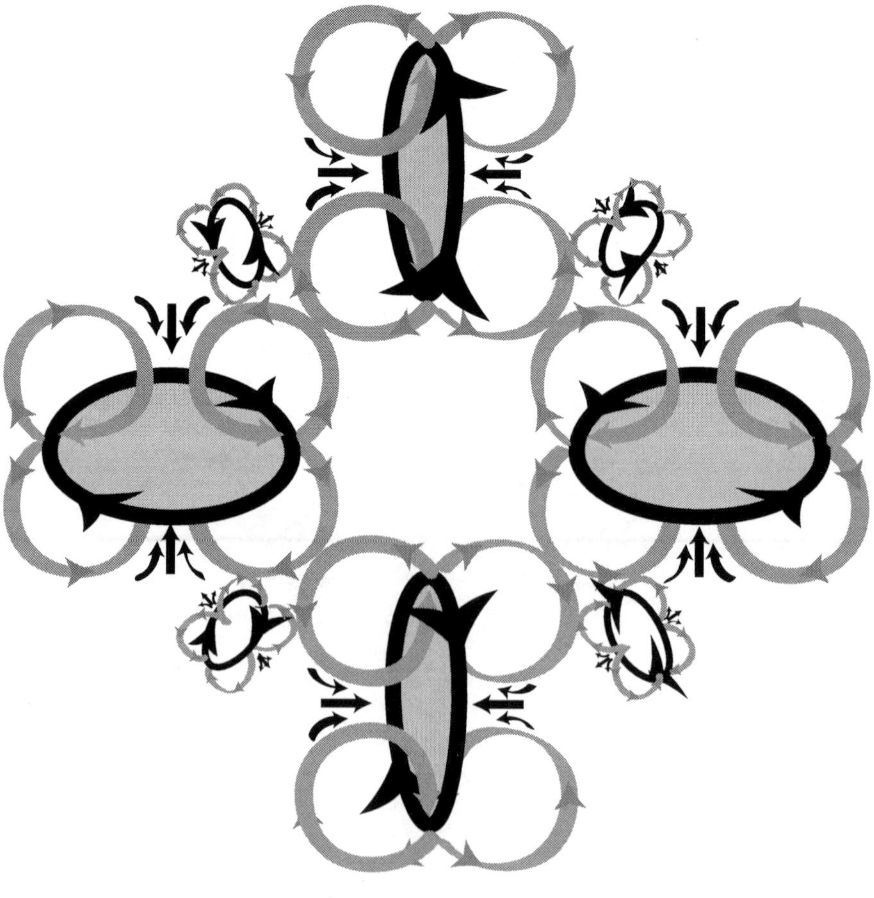

According to *Flow Theory*:

• Low Pressure Spiral Galaxies with Black Holes and High Pressure Lenticular Galaxies may align spins, attract, interact and bend around and bond in loops with dipolar flow and **dynamic interactive stability** to form an Elliptical Galaxy.

• Elliptical Galaxies have similar structures to some of the chemical elements.

Elliptical Galaxy

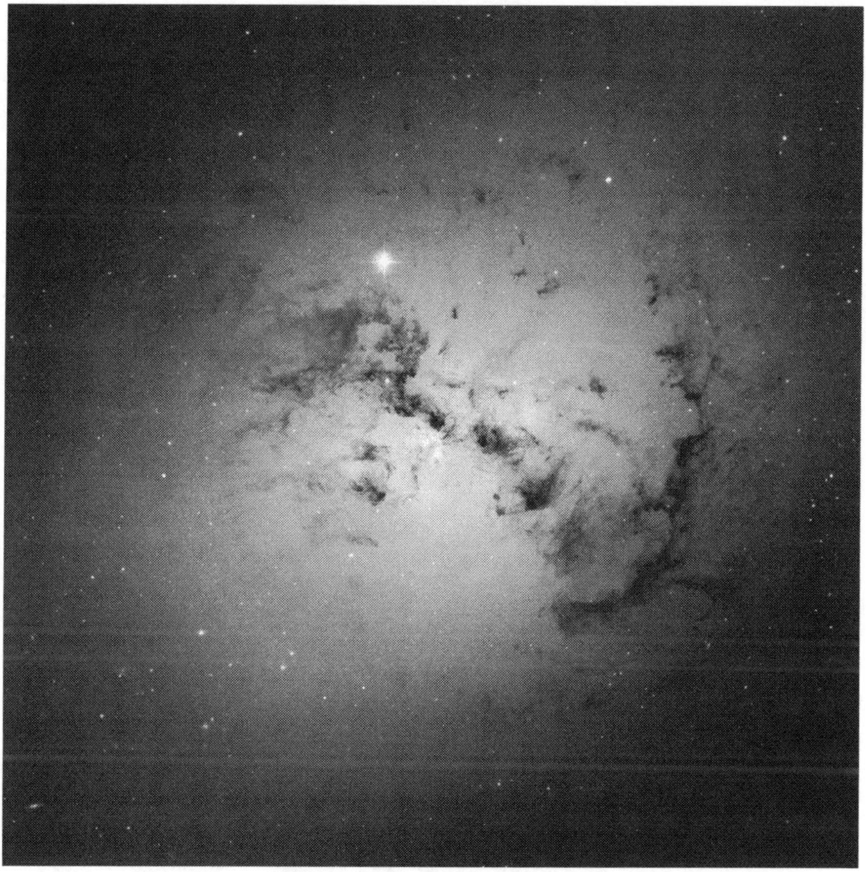

Photo courtesy of ***NASA***
from the ***Hubble News Release Image Archive***

Cake Pan Spin and Objects Experiment

Procedure:

1) Place (without spinning) one or more small objects near the center of a round cake pan.
2) Spin the cake pan with the non-spinning objects in the cake pan.

Diagram:

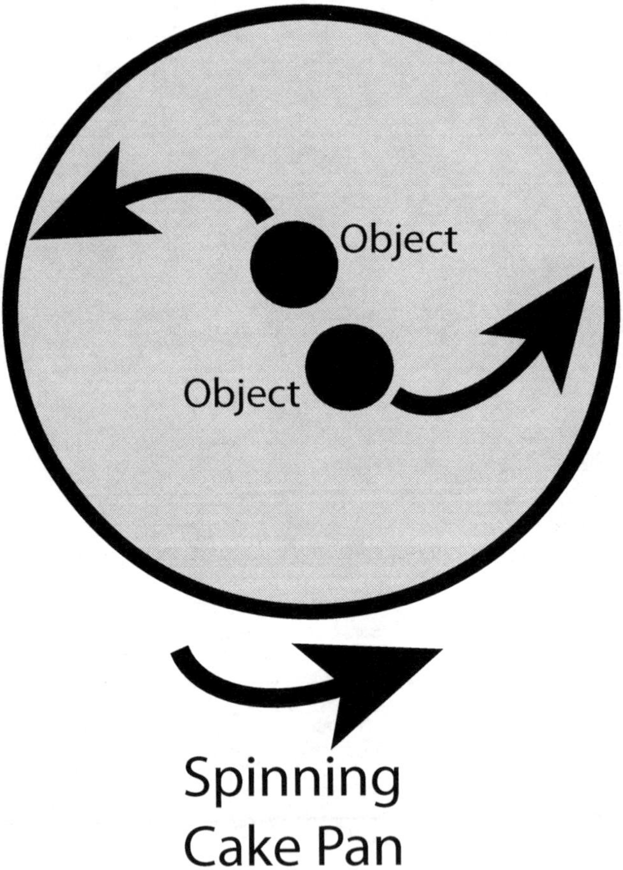

Observations:

• Non-spinning objects inside of a spinning cake pan may move from the center of the cake pan in an arc toward and following the circumference of the spinning cake pan.

Interpretation:

According to *Flow Theory*:

• A circular flow system in an environment of air may form a low pressure center region. The environment inside the circumference of the circular flow system flows toward the circumference by following the inside circular counter flows that are generated by the circumference, and forces are not needed to explain vortex dynamics.

Vortex

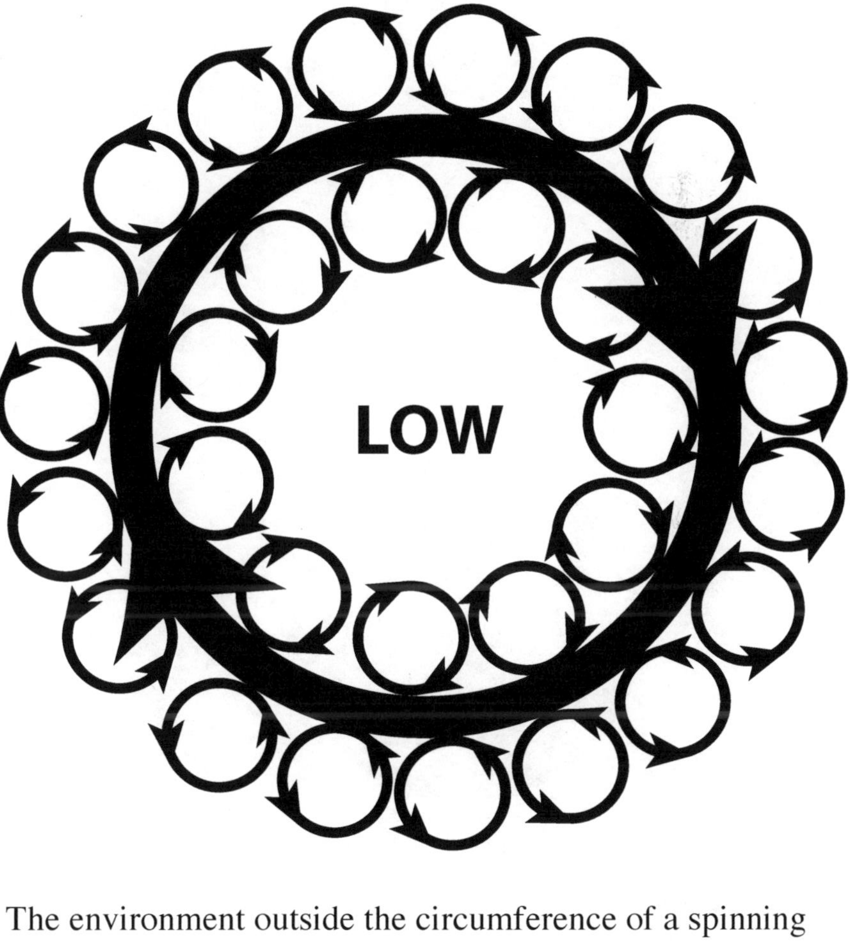

• The environment outside the circumference of a spinning object or circular flow system may follow the outside circular counter flows (generated by the circumference) toward the circumference, and the concept of gravity is not needed to explain surface attraction!

Ball Spin and String Experiment

Procedure:

Spin an extra-large ball, and then slowly bring a long piece of hanging string or thread (non-spinning) close to the side of the spinning ball.

* Use an extra-large ball and a thin piece of string or thread.
* A turntable and bowl may be used to spin the ball.
* Be sure to use types of ball and string that do **not** have electrostatic attraction: leather or rubber ball, cotton thread.

Diagram:

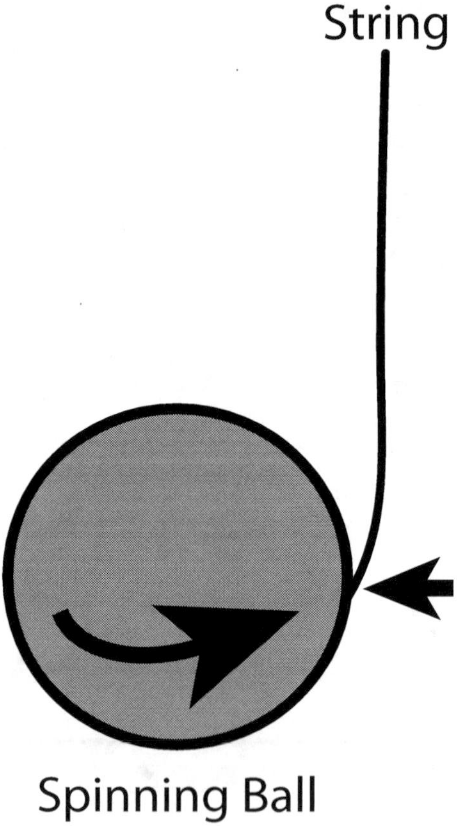

Observations:

• A non-spinning string may move toward the side or surface of a spinning ball such that: the faster the ball is spinning, the faster the string moves toward the surface of the ball.

Note: For the ***Ball Spin and String Experiment***, many factors must be considered when interpreting why the string moves toward a spinning ball: electrostatic attraction between certain types of ball and string combinations, flow in the direction toward the surface of the ball, the velocity of the spinning ball, the distance between the ball and string, the surface texture of the ball, and for the string its mass, density, fray, and material (use 100% cotton thread).

Interpretation:

According to *Flow Theory*:

• The environment outside the circumference of a spinning object or circular flow system may follow the outside circular counter flows (generated by the circumference) toward the circumference to produce surface attraction, and gravity is **not** needed to explain surface attraction!

• Surface attraction is a **dynamic** process that occurs with **spinning** objects such that: the greater the velocity of a spinning object, the greater the surface attraction of the spinning object.

• The atmosphere of a planet may affect its surface attraction (Venus has a high atmospheric pressure (Sparrow) and a high surface attraction for its slow rotation velocity (Sparrow)), and the surface density of a planet may affect the surface attraction (Mercury has a high surface attraction for its small mass (Sparrow)).

• As the Earth spins, the pressures at the fronts of the moving surface structures of the Earth become higher and the pressures at the sides and backs of the moving surface structures become lower. According to the *Laws of Flow* from the book *Flow Theory*, flow occurs in the direction from regions of higher pressure toward regions of lower pressure. Thus, as the entire Earth spins, flows occur in the direction from the higher pressure fronts of moving Earth surface structures toward the lower pressure sides and backs of the moving Earth surface structures to produce surface attraction with overlapping flows toward the entire surface of the Earth.

Resources

Aczel, Amir D. *Pendulum: Léon Foucault and the Triumph of Science*. New York: Washington Square Press, 2003.

Anastopoulos, Charis. *Particle or Wave: The Evolution of the Concept of Matter in Modern Physics*. New Jersey: Princeton University Press, 2008.

Capra, Fritjof. *The Science of Leonardo: Inside the Mind of the Great Genius of the Renaissance*. New York: Anchor Books, 2007.

Haramein, Nassim, presenter. *Black Whole*. DVD. California: GAIAM, 2011.

Harriman, Charles. *Flow Theory*. Ferndale, Washington: Charles Harriman Publishing, 2013.

Laidler, Keith J. *The Harmonious Universe: The Beauty and Unity of Scientific Understanding*. New York: Prometheus Books, 2004.

Longair, Malcolm S. *Our Evolving Universe*. New York: Cambridge University Press, 1996.

Malley, Marjorie C. *Radioactivity: A History of a Mysterious Science*. New York: Oxford University Press, 2011.

Press, H. J. *Science Experiments*. New York: Sterling Publishing, 1998.

Sparrow, Giles. *The Universe and How to See it: A Practical Guide to Viewing and Understanding the Night Sky*. New York: Reader's Digest, 2001.

Made in the USA
Monee, IL
04 November 2026

67352870R00015